龙溪-虹口国家级自然保护区管理局 编著
By The Management Bureau of Longxi-Hongkou National Natural Reserve

A Brief Introduction of Giant Panda

简说大熊猫

四 川 科 学 技 术 出 版 社
Sichuan Publishing House of Science & Technology

大熊猫家园——龙溪-虹口国家级自然保护区
Habitat of Giant Panda—Longxi-Hongkou National Natural Reserve

序

我国改革开放四十多年来，经济迅速发展，人民生活质量日益提高。在这个新时代背景下，政府对生态修复与保护有了高度认识和重视，老百姓也有绿色、生态的强烈愿望和需求。自20世纪90年代实施长江流域“天然林资源保护工程”以来，限制森林砍伐、退耕还林、退耕还草，有效地遏制了森林生态系统的破坏，减少了长江和黄河源头的水土流失，人们的生产和生活环境得到明显改善。

被称为国宝的珍稀动物大熊猫就受益于中国政府强有力的生态保护政策，自1992年“中国保护大熊猫及其栖息地工程”实施以来，无论是就地保护还是迁地保护都取得了显著成就。大熊猫是动物界的“伞护种”，保护了大熊猫和栖息地，也就保护了共同生活在这一区域的各种动物，使生物多样性得到维系，生态系统更完整。

龙溪–虹口国家级自然保护区是距离特大城市成都市最近的一个保护地，区域内地形、地貌复杂，海拔高差大，生物多样性丰富，分布着多种特别珍稀的动植物。自保护区成立以来，认真致力于大熊猫、金丝猴、牛羚等野生动物的监测、巡护、保护以及科学研究与公众的科普教育事业，通过保护区广大员工共同努力与付出，取得了非常优异的保护成效。

龙溪–虹口国家级自然保护区管理局组织编写这本面向青少年的大熊猫知识读物，是强化青少年教育的具体行动，使他们能更好地了解大熊猫，从而参与到保护大熊猫的事业中来。本书语言简练，表述准确，科学性和知识性强，而且图文并茂，图片唯美。青少年朋友们阅读本书之后，会对大熊猫这个珍稀物种有一个轮廓性的了解和认识，这会有利于提升青少年的生态环境保护意识，也会增强保护大熊猫的责任感。总之，本书是一本有益于青少年健康成长和普及生态保护知识的读物，开卷有益，值得一读。

2019年8月

Over the past 40 years of implementing reform and opening-up policy, Chinese economy grows rapidly and people's life quality is improved increasingly. Under the background of this new era, the government has highly recognized and attached great importance to ecological restoration and protection, while people also have a strong desire and demand for green and ecological environment. Since the implementation of the "Natural Forest Resource Protection Project" in the Yangtze river basin in the 1990s, people's production and living environment have been improved significantly by restrictions on deforestation and conversion farmland to forests and grassland that effectively curbed the destruction of forest ecosystem and reduced soil and water loss at the headwaters of the Yangtze River and the Yellow River.

Giant panda, known as the national treasure, has benefited from the strong ecological protection policies of the Chinese government. Since the implementation of Protection of Pandas and Pandas' Habitat Program in 1992, remarkable achievements have been made in both on-site and off-site conservation. The giant panda is an "umbrella specie" in the animal kingdom, which means that protecting the giant panda and its habitat is to protect all kinds of species lived in this area. Thereby the biodiversity is maintained and the ecosystems are more integrated.

Longxi-Hongkou National Natural Reserve is the closest protected area to the mega city of Chengdu, with complex topography, large altitude difference, rich biodiversity and a variety of rare animals and plants. Since the establishment of the Reserve, it has devoted itself to the monitoring, patrolling and protecting of giant pandas, golden monkeys, gnus and other wild animals, as well as scientific research and public science popularization and education. Through the joint efforts of the staff in the Reserve, excellent conservation results have been achieved.

Longxi-Hongkou National Natural Reserve Administration compiles this reading material that is involved with the knowledge about giant pandas for young people, which is to embody the implementation of ecological civilization, to strengthen the youth education action so that teenagers can better understand the giant pandas and to promote them consciously participating in environmental protection and the cause of giant panda protection. This book is concise in language and accurate in description with strong science and knowledge, as well as beautiful pictures. After reading this book, young friends will have a general understanding of giant pandas. This will not only enhance the awareness of ecological protection of teenagers, but also the sense of responsibility to protect giant pandas. In short, this book is good for the healthy growth of young people and the popularization of ecological protection knowledge. Reading is always profitable, besides this book is worth reading.

Hu Jinchu
August 2019

前言

现在，大熊猫热持续升温，圈粉几亿。为了帮助青少年和忙于工作的成人，快速、简单了解可爱的大熊猫，作者尽量把有关大熊猫的知识浓缩和简化，并配上有趣的图片，便于读者轮廓性地了解大熊猫、靠近大熊猫。在这个信息量非常大的时代，作者认为阅读一本书之后，能得到一些启发和认识就相当不错了。本书的创作团队都在保护大熊猫等野生动物的一线工作，他们已把对野生动物的热爱和保护的责任感深深地印记在内心。通过理论与实践，对大熊猫的了解、认识更为深刻。本书把大熊猫与其他野生动物不同之处提取出来，共分为六部分进行概要阐述，作者希望人们通过阅读本书，获得正确的大熊猫知识和信息，提高大熊猫保护意识。从而在实际工作、学习及生活过程中，践行爱护环境，保护大熊猫等野生动物的理念，养成珍惜生命、爱护动物、节约能源的良好习惯，为人类和可爱的大熊猫留住绿水青山、共享蓝天白云做出自己的贡献。

Today, the fever of giant pandas continues to heat up, winning countless fans. The authors try their best to condense and simplify the knowledge of giant panda with interesting pictures in order to help teenagers and busy adults to quickly and simply understand the lovely giant pandas, so that they are convenient to understand the giant pandas and get close to them in generally. In this age of large amounts of information, the authors think it is quite good to get some inspiration and knowledge after reading a book. The writers of this book are all working on the front line of protecting pandas and other wild animals. Their love for wild animals and sense of protection responsibility are deeply imprinted on their hearts. Through theory and practice, the understanding of the giant panda is more profound. This book extracts the differences between giant pandas and other wild animals, which are summarized in six parts. The authors hope that people can get the correct knowledge and information of giant pandas through reading this book and improve their awareness of giant pandas' protection. Therefore, people can practice the protection concepts such as protection environment, giant pandas and other wild animals during the processes of their actual work, study and life, forming good habits of cherishing life, caring for animals and saving energy. Besides, people should make their own contributions for mankind and the lovely pandas to keep green mountains and clear waters, shared blue sky and white clouds.

目录

大熊猫的历史与分布

The History and Distribution of Giant Panda

始熊猫化石发现于云南省禄丰县、元谋县，地质年代为中新世晚期，约为800万年前。因此，现代的大熊猫是由始熊猫进化而来，属剑齿象古生物群，同时期的动物多数已经绝灭，大熊猫通过自身变化去适应环境的变化而生存下来，所以大熊猫成为一个古老的活化石物种。

The fossil evidence of the primeval panda (Ailuaractos lufengensis) was found at Lufeng and Yuanmou of Yunnan Province in the late Miocene, about 8 million years ago. Therefore, the modern giant panda is evolved from the primeval panda, and belong to da-Stegodon Fauna. Most of the animals in the same period have been extinct, however, giant panda survived. Through their own adaption for the environment change, giant panda was called living fossil.

中国古代就有大熊猫的记载，称谓有貔貅、貘、食铁兽等。几千年来一直流传着伴随人类社会进程的很多故事，如古代在战场上正在冲锋拼杀之时，如果出现高举的大熊猫旗帜，敌对双方会立即停止战争；老百姓把大熊猫皮张挂在屋里，用以驱邪避鬼；唐代，武则天曾将大熊猫和大熊猫皮赠送给当时的日本天皇，以示友好。

Panda was called pixiu, mo, and iron eating beast in ancient China thousands of years, many panda stories have been spread along with the process of human society. For example, in ancient times, when the battle is charging or fighting, if the giant panda flag was raised, the enemy sides will stop fighting immediately. People also hung panda skin in their house to ward off evils. During the Tang Dynasty, Emperor Wu Zetian gave pandas and panda skins to the emperor of Japan as a gesture of friendship.

1869年，法国传教士戴维在四川省宝兴县获得大熊猫标本，在与巴黎自然博物馆馆长米勒·爱德华研究后，定名为大熊猫（拉丁文学名*Ailuopoda melanoleuca*）。当地老百姓称大熊猫为白熊、竹熊、华熊、花熊、银狗等，中国近代称大熊猫为猫熊、大猫熊。由于语言习惯，现在统称为大熊猫。

After the Opium War, in 1869, French missionary Armand David obtained giant panda specimens at Baoxing, Sichuan Province. Went through study with Paris Museum of Nature curator Miller Edward, named as the giant panda (Ailuopoda melanoleuca). People called giant pandas as white, bamboo, Chinese, and color bear or silver dog. In modern times, giant pandas were called giant panda.

古时候，大熊猫分布较为广泛，北至北京周口店，南到东南亚的缅甸、越南、老挝等国。现在，大熊猫仅分布于中国境内。主要分布于四川、陕西、甘肃的岷山、邛崃山、秦岭、凉山、大小相岭等山系。大熊猫分布的五大山系，由于人类的生产活动，多被切割成互不相连的块状，呈孤岛化。

In ancient times, giant pandas were widely distributed, north to Zhoukoudian Beijing, and south to Myanmar, Vietnam, Laos and other countries in Southeast Asia. Now, giant pandas are only found in China, mainly distributing in Min, Qionglai, Qinling, Liang, Xiangling and other Mountains at Sichuan, Shaanxi and Gansu Province. Unfortunately, due to human activities, these five mountain systems are cut into unconnected blocks, isolating from each other.

大熊猫独特的生理与习性

The Unique Habit and Physiology of Giant Panda

在动物分类上，大熊猫属于食肉动物。为了生存，由食肉到杂食，再逐渐过渡到主食竹子，竹子占其食物总量的99%。大熊猫仍然留存着食肉动物的体型结构和较短的消化道。成年大熊猫每天食竹量为10~20千克，6~8小时就会排泄出来，多食快排，以适应低能量食物。

In terms of animal classification, giant panda is a carnivore. For surviving, giant panda changed from carnivorous to omnivorous, and then gradually to staple bamboo. But giant panda still retains the body structure as a carnivore with a short digestive tract. Adult giant panda can eat about 10-20 kilograms bamboo per day. The excreting time is around 6-8 hours. This is their strategy to adapt to low energy food.

大熊猫多为独居，最大活动范围为30平方千米左右。根据不同山系的竹子丰富度和不同地形，其分布密度有所变化。大熊猫喜欢居住在竹子种类多、林下竹子密度大、背风面、有水源、山体坡度在20度~30度、生长有一定高大乔木的地域。

Giant pandas mostly live alone, the maximum activity range is about 30 km^2. According to the bamboo varies and the abundance of topography in different mountain systems, giant panda has different population density. They prefer to live in the area where there are different kinds of bamboos under the forests with high density. And where has leeward, water source, slope of mountains at 20-30 degrees with tall trees.

现在，大熊猫大多生活在青藏高原的东南缘，大熊猫栖息地的生态系统比较完整，环境洁净。大熊猫喜欢饮水，而且是畅饮，特别喜欢洁净的山泉水。兴致大发时，还会在山泉边或河边戏水、沐浴。

Most of the giant pandas are living in the southeastern edge of the Qinghai-Tibet Plateau, where the ecosystem of giant panda habitat is relatively integrated. Giant pandas like drinking water, especially like the clean mountain spring water. When giant pandas are full of enthusiasm, they will bath and splash at the spring water or river.

大熊猫善于爬树，一岁之前多攀爬在树上，这样在妈妈觅食回来之前比较安全。大熊猫具有强有力的前肢，其指尖锐有力，还可以伸缩，有利于攀爬树木。由于大熊猫在森林里生活，常年穿梭于林间，因此，为了适应环境，大熊猫四肢行走为内“八”字形，身体常常左甩右晃。

Giant panda is good at climbing trees. Their cubs speed most of their times staying in the tree, which is a relative safe place when their mothers are foraging at another place, before one years old. Giant pandas have powerful sharp forelimbs and powerful fingers that can be telescoped to help them climb trees. Giant pandas spend most of their time in the forest because they have to shuttle between the woods. Therefore, giant pandas walk as pigeon toe shape to adapt this circumstance. That is why the body of giant panda likes swaying when they are walking.

大熊猫的视力相对较差，这与大熊猫的生活环境有关。在林间觅食竹子，不需要敏锐的视力。但其嗅觉和听觉非常发达，敏锐的嗅觉有助于辨别竹子的品质和识别其他大熊猫的气味标记信息；很好的听力有利于识别天敌，回避威胁，也利于监听同类的呼唤及警告。

The vision of the giant panda is relatively poor, which is due to the environment of giant panda habitat. Foraging for bamboo in the forest, sharp vision is not required. But their sense of smell and hearing is developed. Keen sense of smell can be used to identify the quality of bamboo and identify odor marker information of other giant pandas. Good hearing ability helps them identify predators, avoid threats and listen calls or warnings from congeners.

大熊猫除了正常的五指外，为了适应进食竹子方便，又进化出第六指，也称伪拇指。这个伪拇指起的作用就非常大了，当大熊猫用尖利的牙齿咬断一根一根竹子以后，就靠伪拇指与其他五指配合，可以游刃有余地拿握竹子。

Giant panda evolved the 6th finger, as known as false thumb which works perfectly, in order to eat bamboo. After giant panda bites bamboo with its sharp teeth, the false thumb and other 5 fingers can handle bamboo easily.

为了进食坚硬的竹子，大熊猫的头骨逐渐变大、咀嚼肌也变得强劲有力，可以毫不费力地咬断坚硬的竹子。要把咬断成节的竹子研磨碎，还要靠大熊猫发达而宽大的臼齿。因为大熊猫的祖先是食肉动物，所以其齿式与起研磨作用的臼齿也必须适应食竹的需求。

Bamboos are solid. However, giant panda has a large skull with powerful chewing muscles that can easily break through the "tough" bamboo. And the wider-spread molars can grind up the bamboo pieces. Because giant panda was a carnivore, its teeth and grinding molars had to change to adapt the bamboo diet.

大熊猫一般以坐姿进食，有时也仰卧或侧卧进食。在进食竹秆时，它会用前肢与牙齿配合，把坚硬的竹秆外皮剥干净，然后咬断成节进食。在吃竹叶时，它会用嘴从竹秆上把竹叶咬下来，并集中起来，用前肢拿握成把，左边一口、右边一口地进食，这就均衡了左右牙齿的磨损，相当科学、合理。

Generally, giant pandas eat in a sitting position, sometimes lying on its back or side. When they eating bamboo poles, they will use their forelimbs and teeth to peel the skin of the hard bamboo pole, and then bite into knots to eat. When they eat bamboo leaves, firstly they will use their mouth to bite off the bamboo leaves from the bamboo pole, then gather leaves together and hold leaves with their forelimbs, finally bite with both sides' teeth alternate. This strategy equalizes the wear, left tear and right teeth, which is quite scientific and reasonable.

大熊猫四肢粗壮，走起路来呈典型的内“八”字步，这绝对不是人们印象中的畸形。大熊猫的生活地域大部分都是森林，林下竹子茂密，长期在这种环境中穿梭走动，最便利的行走方式就是把四肢内撇。在科学家看来，这也是一种对环境的适应。内“八”字步配上大熊猫圆胖的身体，应该是非常协调的，看起来也很有趣。

The limbs of giant panda are strong, showing pigeon toe characters. This is definitely not deformity. Giant pandas live in forest areas, most of forest area is with dense bamboo undergrowth. For a long time, they move around in this environment. The most convenient way is to turn their limbs intorted. For scientific perspective, it is also an adaptation. The pigeon toe with the chubby body of giant panda looks integrated and interesting.

大熊猫通常在每年春季的2~5月份发情、配种，发情高潮期仅有3天左右。大熊猫一年只发情一次，个别的偶尔在秋季发情。在春季发情季节，几只雄兽会聚集到发情雌兽的生活领域，展开交配权的争夺，但时间很短，大约为一周。

In the spring of the year from February to May is the mating season of giant panda, but the height of the rutting are only about three days. Pandas are only in estrus once a year. Few induvial may rut in autumn. During the spring rutting season, several males gather around to compete for mating rights in the area of the female active. But it is a short time, only about a week.

大熊猫的婚姻是比较挑剔的，雌雄双方都要经过严格的挑选，有一方不满意，都不会进入交配阶段。交配一般在地面进行，个别在树上交配，雌兽可以接受多只中意的雄兽交配。

The “marriage” of giant panda is nitpicky; both male and female have to go through strict selection. If one side is not satisfied, they will not go into the mating stage. Mating is usually done on the ground, occasionally on trees, and females can mate with several satisfied males.

在人工圈养条件下，为了充分利用育龄期大熊猫的资源，提高繁殖效率，不能都让大熊猫自由恋爱，往往人为安排配对。科技人员一般会将有交配经验的雌性大熊猫与初次参与配种的雄性进行配对；反之将有交配经验的雄性大熊猫安排给初次配种的雌性，另外还可以将处于发情高潮期的雌性或雄性进行调包，这都是人工干预大熊猫配种的重要手段。

In captivity, people in order to make full use of the breeding age of giant panda resources, improve the efficiency of reproduction, people cannot ensure all giant pandas having free love. Often go through artificial arrangements for pairing. In general, scientists will pair the female pandas with mating experience and the males that participate in mating for the first time. On the contrary, the males with mating experience will be arranged for the females that mate for the first time. In addition, the females or males who are in the peak of estrus also can be exchange pairing.

大熊猫雌兽的怀孕期差别非常大，怀孕期一般从80~200多天不等，预产期非常难以判断。学者认为，大熊猫胚胎存在游离和延迟着床的现象，当外界环境和身体状况良好的时候，胚胎才能着床。

The pregnancy period of mother giant panda is flexible, generally ranging is from 80 to over 200 days. Therefore, the due date is very difficult to estimate. Scientist believes that the embryos of giant panda may have free and delayed implantation until the external environment and physical condition get suitable, then the embryo implantation.

在野外，大熊猫多为单胎，双胞胎降生概率为20%~30%。在人工圈养条件下，通过自然交配与人工授精相结合，双胞胎的出生率能达到50%左右，不仅如此，人工圈养大熊猫还偶有三胎出生。

In the wild, giant pandas usually have singletons. The birth probability of twins is about 20%-30%. In captivity, under natural mating and artificial insemination conditions, the twin birth rate can reach about 50%, sometimes may get triplets.

大熊猫刚出生时很多器官发育不全，通体肉红，被毛稀疏。初生幼子没有睁眼，也没有听力，但能叫，声音很大，有吸吮母乳的本能。一周之后，被毛和器官才有所变化。大熊猫幼子在高蛋白、高脂肪、高能量的母乳喂养下，生长发育迅速。

The organs of newborn giant panda are not fully developed. The whole body shows red with sparse hair. And the eyes are close, also no hearing ability, but they can cry loudly and have the instinct to receive breast milk. The appearance of hair and organ will start change after a week. Newborn giant panda cubs grow and develop very fast by having high protein, high fat and high-energy breast mike.

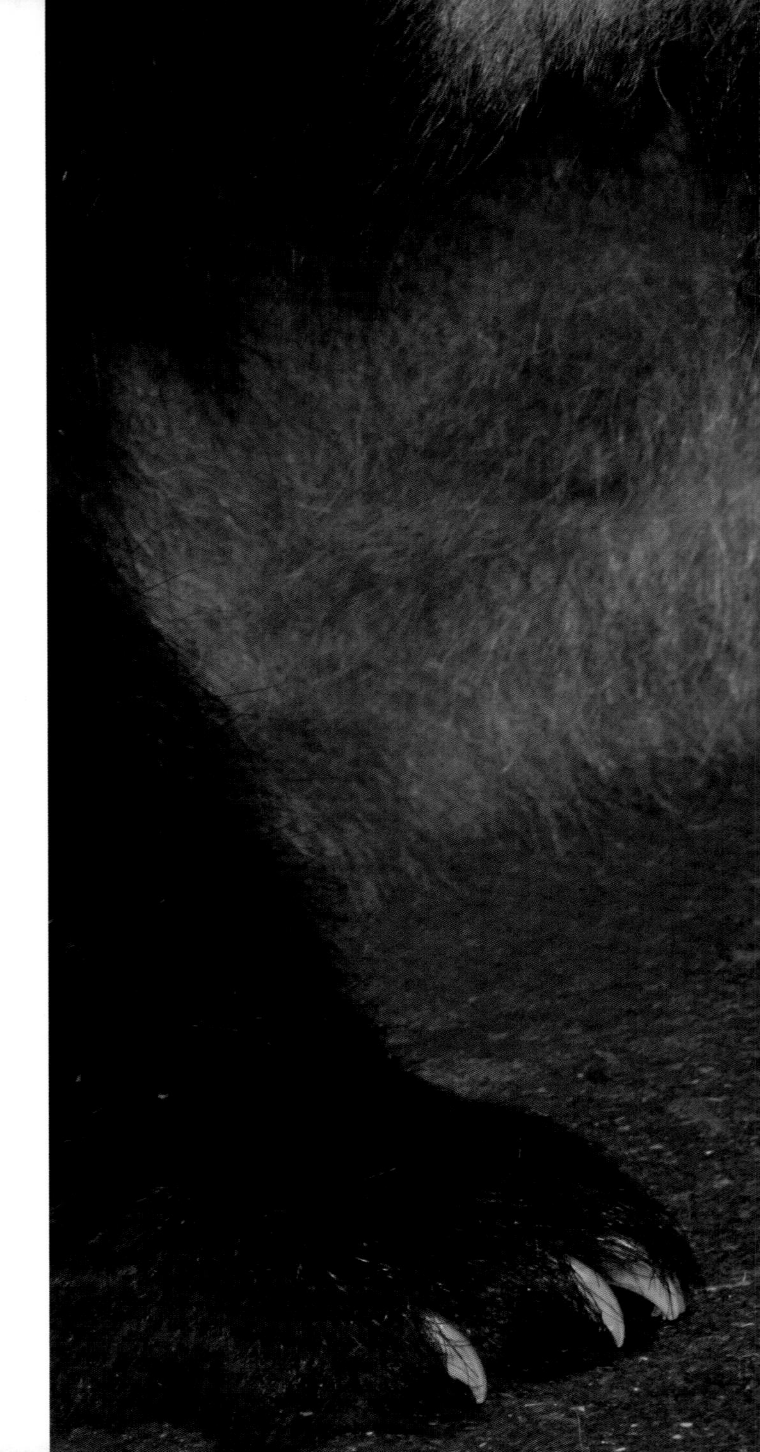

魏荣平 摄

大熊猫初生幼子的体重仅为母体的千分之一，80~160克。成年大熊猫体重为80~120公斤，雄性个体大，雌性稍小。幼子在成长过程中，会跟着母亲学习很多东西，野外生活的大熊猫在两岁以后离开母亲独立生活，人工圈养的到1岁时就会与母亲隔离。

The newborn giant panda cub weight is only one thousandth of the mother, about 80-160 grams. The weight of adult giant panda is about 80-120 kg. The body size of male is a little larger than female. The cubs learn a lot from their mother during growing up. In the wild, they will leave their mother to live independently after 2 years old. In captivity, the cubs will be separated from their mother after 1 years old.

刚出生的大熊猫幼子长着一条长尾巴，约占体长的三分之一。随着幼子的生长发育，尾巴停止生长发育，到成年后只剩下10厘米左右的短尾了。初生幼子的长尾巴也印证了大熊猫祖先具有食肉动物的身体特征。

The newborn giant panda cub has a long tail, about one-third of the body length. As cubs growing and developing, the tail stops growing and only maintains about 10cm at adulthood. The long tail of newborn cub also proves that the ancestors of giant panda had the physical characteristics of carnivores.

大熊猫雌兽生下幼子后，会全身心地投入到抚育幼子中，产子后的三天里大熊猫雌兽几乎不吃不喝。因为初生的大熊猫幼子发育不全，身体很脆弱，需要母亲的温暖和照顾。大熊猫雄兽配种成功以后就离开雌兽的领地，它是不参与守护幼子的，如果雌兽离开幼子太久，被伤害的概率会大大增高。

After giving birth to a cub, the mother giant panda pays all attention to raising the cub, at the first three days almost do not eat or drink. Because the newborn giant panda cub underdevelopment and the body are very fragile, they need their mother's warmth and care. Male giant pandas will leave the territory of the female giant panda after mating. They are not involved in guarding the cubs. Therefore, if the mother left cubs for too long, the risk of cubs would be greatly increased.

很多大熊猫雌兽会做出弃子行为，特别是第一次产子的雌兽，对突然生下的幼子无所适从，显得特别慌乱，不知道怎样处理。个别的雌兽会被刚生下的幼子吓一跳，并把幼子抛到一边。还有就是双胞胎降生时，大熊猫雌兽都是用嘴含幼子，没有抚育经验的雌兽不懂得如何同时照顾两只幼子，因此，雌兽会遗弃一只小而体弱的幼子。

Some of mother giant pandas have the phenomenon of abandoning cubs, especially the first time to be a mother. Suddenly give birth to a baby makes them nothing to follow, fluster and do not know how to deal with. Some mothers will be startled by the newborn cubs and throw them aside. What's more, when twins are born, mother giant pandas are holding them by using their mouth. The inexperienced mother giant pandas do not know how to take care of two cubs at the same time, therefore, mother giant pandas will abandon one that is small and weak.

大熊猫深受人们的喜爱，因此大家都很关心它的寿命。其实大熊猫寿命与它生存的环境条件密切相关，最重要的是食物，能够采食到优良、充足的竹子，大熊猫的体况就好。栖息地中没有天敌和疾病的威胁，没有人类的干扰，也会使大熊猫寿命延长。另外，遗传因素也左右着大熊猫寿命的长短。总之，野外大熊猫寿命在18岁左右，人工圈养的在25~30岁。

Giant panda is quite popular around the world; everyone is concerned about its lifetime. In fact, the life span of giant panda is closely related to its living environment. The most important fact is food, whether they have enough high-quality bamboo to eat, and will determine the physical condition of giant panda. The absence of natural predators, diseases and human interference in their habitats could also prolong their lives. Of course, genetic factors also play an important role. Overall, wild giant pandas will live around 18 years old, and giant pandas in captivity will live between 25 to 30 years old.

在动物园里参观大熊猫，多数时候都见大熊猫身稳不动，觉得大熊猫比较懒。其实不然，大熊猫在野外走动觅食的时间很长，活动时间在14个小时以上。每天活动频率较高的时段是在早上的4~6点和下午的5~7点。生活在动物园里的大熊猫，由于人为给食，因此活动较少。另外，竹子营养成分少，减少能量消耗也是一种生存策略。

Most of the time giant panda does not move a lot while you are visiting a park or zoo. You may think the giant panda is lazy. Actually not, giant panda will spend much time ambulating and foraging in the wild. The activity time is more than 14 hours a day. The periods with high activity frequency are from 4 to 6 am and from 5 to 7 pm. The captive giant pandas are fed by humans, so they having fewer active times. In addition, reduce energy consumption is a survival strategy for low nutrition of bamboo.

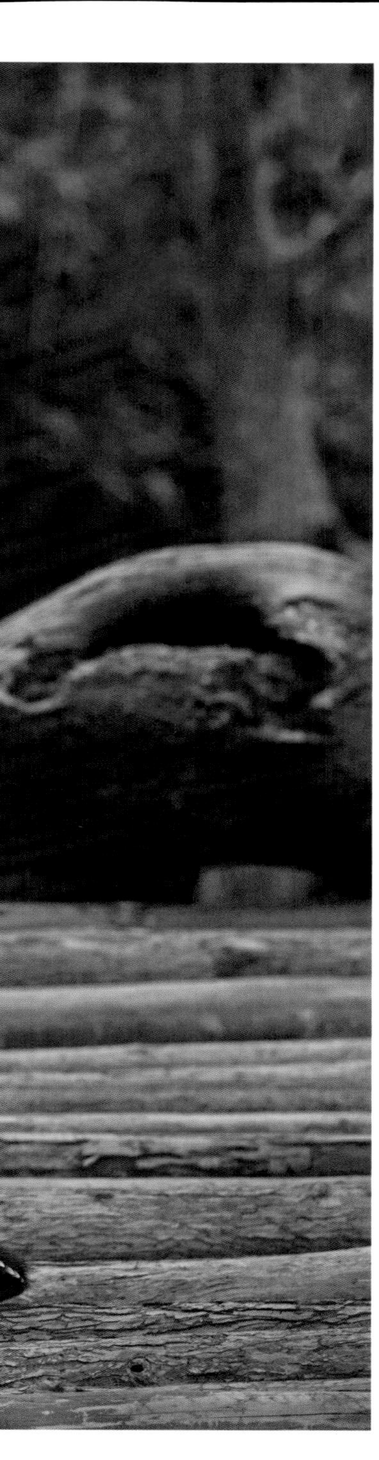

大熊猫怕热不怕冷。大熊猫皮毛保暖性好，皮毛下还有一层厚厚的脂肪，能够在零下十几度的环境中生活。大熊猫汗腺不发达，几乎没有汗腺，身体散热差，所以当环境温度超过25℃时，大熊猫会感到很难受。

Giant pandas are afraid of heat, but not afraid of cold. The fur of giant panda is good at keeping warm and has a thick layer of fat under it, which allow them to live under minus 10℃ degree. The sweat glands of giant panda are not developed. Therefore, the thermolysis of giant panda is poor. They will be very uncomfortable if the temperature rises above 25℃ degrees.

大熊猫在进化过程中，为了便于采食竹子，其前掌进化出伪拇指，形似一个肉垫，这就非常方便拿握和采食竹子；为了适应主食竹子，大熊猫的头骨、嚼肌和臼齿都发生了适应性变化，咬力和臼齿研磨力大大增强。

In the history of giant panda evolution, its forelimb evolved a false thumb for their bamboo diet. False thumb likes a meat pad, which is very convenient to grasp and eat bamboo. In order to adapt the bamboo diet, the skull, chewing muscles and molars of giant panda had specialized with increased biting and grinding force of molars.

大熊猫每天进食大量的竹叶和竹秆，竹子主要由坚硬的木质素与纤维素构成，大熊猫通过在口腔中咀嚼、物理粉碎，最后还是有2~3厘米长，这对大熊猫的消化道有一定损伤。但大熊猫的消化道会分泌大量的黏液，通过包裹食物来保护消化道。在人工饲养下，会有1~2天的排黏日，大熊猫会趴卧不动，有明显的腹痛症状，待排除黏液后才会恢复正常活动。

Giant pandas eat a lot of bamboo leaves and bamboo rods every day. Bamboo is mainly composed of hard lignin and cellulose. Through chewing in the mouth with physical crushing, the length of bamboo pieces still has 2-3cm, which could damage the digestive tract of giant panda. However, the digestive tract of giant pandas will secrete a lot of mucus, which can coat their food and protect their digestive tracts. In captivity, there are 1 or 2 days, giant panda will discharge mucus for a period of time. They will lie down and show motionless with abdominal pain. After finish discharging mucus, giant panda will return to normal activities.

大熊猫喜好独自游荡，性情孤僻，随意性强。它成天在自己的领地中寻找可口的竹笋和竹子，吃饱就睡，醒了就游山玩水，又没有其他动物威胁，日子过得非常悠闲自在。它有领地意识，但在其活动地域内不固定巢穴和住处，边吃、边走、边休息，随遇而安。

Giant pandas like to wander alone, they are unsociable and free rein. They are looking for delicious bamboos in their own territory every day, sleep after full, sightseeing among hills and up or down rivers after wakes up. There is no threat from other animals for adult giant pandas. Giant pandas have territorial consciousness, but there is no fixed nest or residence place in their activity area.

大熊猫虽然走上了素食主义的道路，在它的食谱中，竹子占据95%~99%，然而个别时候还是想吃点荤的食物。在分布有大熊猫的地区，特别是在冬季，发现有大熊猫进入羊圈吃羊的现象，还在野外发现过大熊猫啃食其他动物尸骨的情况，此外，它还对烤过的兽骨特别感兴趣。

Although giant panda has gone to be vegetarian, 95%-99% of its diet is bamboo; occasionally they still want to eat some meat. At the giant panda distribution areas, especially in winter, giant pandas were found eating sheep in the pens. It has also been found that giant panda eating other animal's bones in the wild, and they are particularly interested in roasted bones.

如果一个物种处于濒危状态下，人类必须进行干预，以辅助其种群数量增长，人工圈养大熊猫的迁地保护就是为了这一目的，但一个物种总不能一直生活在城市里或人工饲育。在大熊猫种群达到一定数量之后，野化放归就摆上了议事日程，人工圈养的大熊猫在经过野化训练、恢复本能后，就可以放回到它的家乡去，起到复壮野生种群的作用，使它们在大自然中生存、繁衍，这也是大熊猫保护的最终目标。

When a species is endangered, for conservation purpose, human must intervene to help its population grow. The captive giant pandas of off-site conservation are for this purpose. But a species cannot always live in a city or be raised artificially. After reaching a certain number, the reintroduction project has to be put on the agenda. Captive giant pandas can be released back to their nature habitat after training and restoring instincts and skills, which can strengthen the wild population an helping them survive and reproduce in the wild. This is the ultimate goal of giant panda conservation.

大熊猫的可爱之处

The Lovely Appearance and Behavior of Giant Panda

大熊猫具有黑白两色的体毛特征，四肢、耳朵、眼部为黑色，其他部位的体毛为白色，色差极为明显。在五颜六色的世界里，再也没有什么动物具有这么独特的外貌了，黑白两色极其简单、奢侈、高贵……但自然界物种的基因变异是常事，在大熊猫中也存在棕色和白色被毛的个体。

Giant panda has black and white body. Limbs, ears and eyes are black, other parts of the body is white. Color aberration is very obvious. In this colorful world, no animals have such a unique appearance, black and white are extremely simple, luxury and noble. In the nature, genetic variation is common. Same on giant pandas, they also have brown and white individuals.

在动物世界里，不是具有凶猛的攻击性，就是具备极快的逃匿性；毛皮不是具有警告性，就是具备拟态隐蔽性。大部分动物都不容易靠近，始终与人类保持一定的距离。而大熊猫体态圆润、生性温柔、落落大方、不神经质，使人类产生一种亲近感和靠近的欲望。

In the animal kingdom, some are fierce and offensive, some are fast and good at escape; the fur is either warning or mimicry. Most animals are not easy to get close to, and they always keep a certain distance from humans. But the appearance of giant panda is chubby, gentle in nature, generous and not neurotic, which produces a kind of affinity to human.

大熊猫圆润的头部，两只小眼睛闪闪发光，自然描绘的大眼圈又与两只毛茸茸的耳朵相呼应，大小、布局协调而均衡。大熊猫圆滚滚的躯体，在以内“八”字左右晃动缓慢奔跑时，显得特别可爱。大熊猫对周边环境物体和其他动物都保持着高度的好奇心，喜欢探究和嬉戏。

The roundish head, two small sparkling eyes, big eye sockets and two furry ears of giant panda are integrated. The chubby body running in pigeon toe style appears to be particularly cute. Giant pandas are highly curious about their surroundings and other animals. They are willing to explore and play.

大熊猫不仅拥有落落大方的气质，而且具有随意的自然属性，特别是休息时的姿势充满惬意性，令人羡慕。大熊猫趴卧时，四肢舒展；趴在树上或栖架上时，四肢会自然垂吊；有时会四肢朝天仰卧；有的时候还用前肢蒙上眼睛，或一个后肢搭在树的枝干上；坐姿时，往往背靠树干或石头，省力、舒适。

Giant panda not only has generous temperament, but also has a free rein character. Especially the comfortable rest posture, enviably. When a giant panda lies on its stomach, the limbs are stretched. When lying on a tree or perch, the limbs will hang naturally. Sometimes they would lie on their back with their arms outstretched, or cover their eyes with forelimbs, or put a hind limb on the branches of a tree. When giant pandas are in sitting posture, they often rely on a tree trunk or a stone which is save effort and comfortable.

年幼大熊猫在一起相处时，大熊猫的个体行为非常绅士与优雅。几只幼体大熊猫在一起喝奶时，一般都会各自饮用。有时，喝得快的大熊猫会去抢相邻大熊猫的食物，被抢的大熊猫一般都会发出轻轻警告的叫声，然后用前肢轻轻推开对方，绝对不会发生激烈打斗的行为。个别意犹未尽的大熊猫，还会抓过对方沾有奶的前肢舔食，被舔者也很乐意满足对方的要求。

The individual behavior of giant panda cubs is very gentleman and elegant when they are together. When several cubs drink milk together, they usually drink separately with their own portion. Sometimes, the fast eating cubs will snatch the food of the nearby cubs. The cubs which be robbed will send out light warning yelling, then gently push intruder with forelimbs. Absolutely they will not have fierce fight. If one cub is not satisfied, it will grasp patterner's forelimbs and lick the milk left on.

大熊猫的很多行为动作可爱而滑稽，兴奋的时候会扭动着胖乎乎的身体狂奔，个别时候嘴上会含着一根竹秆嬉戏，还有上树、前滚翻、后滚翻、拥抱、拉拽、相互追逐等。幼年期的大熊猫在一起时，会相互轻轻撕咬，其中一只发出痛疼的轻微叫声后，另外一只会立即终止啃咬。幼年大熊猫成功复制了人类儿童的天真、好奇、幼稚、呆萌、任性、友好的天性，因此会引起人类的内心共鸣。

The behavior of giant panda is cute and funny. They will wiggle chubby body running or contain a bamboo pole in their mouth playing when they are exciting. Sometimes they will climb trees, roll forward, roll back, hug, pull, chase, etc. At juvenile stage, young cubs may bite each other gently. Once, one cub gives a slight yelling of pain, the other one will stop biting immediately. Young giant panda cubs successfully copied the innocent, curious, naive, cute, wayward and friendly from our childhood, this is why giant panda cubs arouse the inner resonance of us.

大熊猫的吃相很有意思，无论吃竹子还是进食窝头和苹果，都会吃出声来，看起来吃得很香，津津有味。吃竹秆时，大熊猫会用牙齿把竹皮剥得干干净净，然后一节节咬断来吃。吃竹叶的时候，大熊猫会用嘴把竹叶理成一把，而后左一口右一口把竹叶咬断进食，从而保持了左右牙齿的均衡使用。

Giant panda's eating habits are very interesting. No matter eat bamboo or eat steamed bread and apples, they will eat out loud and with appetite. When they are eating bamboo poles, giant pandas will use their teeth to peel and clean the bamboo skin first, and then bite and eat. When they are eating bamboo leaves, giant pandas will use their mouth to cut bamboo leaves into a handful, and then bite off the leaves by using both sides teeth alternately. Thereby, it can keep the usage balance of left and right teeth.

大熊猫在哺乳动物中的智商属中上水平，它们为了翻越围墙会爬向高处，还会踩着伙伴身体攀爬，偶尔会成功外逃。还有个别大熊猫会爬到外倾的大树上，在树的上部使劲晃动，让树干倾倒于外侧，它就可以借机在外边溜达一圈。

In mammals, the IQ of giant panda is above the average level. They know climb on a high place to get over walls. They may use partners' bodies to climb and escape, occasionally successed. Some individuals may climb on to the oblique tree, shake on the top of the tree, and let the trunk topple over to escape.

在人工饲养条件下，大熊猫幼子还有一个“人来疯”的特点，当饲养人员进入它的活动场地时，大熊猫幼子会主动跑过来找你玩耍。往往会抱着饲养人员的腿不放，或轻轻撕咬、嬉戏。大熊猫幼子在室外玩耍时，如果周边观赏的游人很多，也会激起它的表演欲望，做翻滚、奔跑、爬树等动作，很讨人喜欢。

In captivity, giant panda cubs also have an “act up” characteristic. When the keeper enters their activity area, the cubs will come and ask for playing together. They often hold the keeper's legs and bite gently for having fun. In outdoor enclosure, giant panda cubs will arouse their desire to perform if they are surrounding with visitors. They will do tumbling, running, climbing or other movements.

无论什么人种，无论什么国籍，为什么男女老少都那么喜爱大熊猫呢？这除了大熊猫独特的相貌外，最主要的还是大熊猫很多行为与人类具有趋同性——温柔、善良、友好、合作、分享、天性。有相同兴趣、有共同爱好、友好相处的人类或动物就容易相互走近，这包括感情上的和身体上的靠近。

Why people all love giant pandas? Giant pandas not only have the unique appearance, but also have the gentle, kind, friendly, cooperative and sharing nature as us. Humans or animals that share the same interests and hobbies tend to attract each other, both in emotionally and physically.

大熊猫的保护

The Conservation of Giant Panda

威胁大熊猫生存的因素主要来自人类的威胁，如人类的各种生产活动：森林砍伐、农田开垦、道路和桥梁的修建、水电站建设等，都会破坏和切割大熊猫的栖息地，从而缩小大熊猫的生存空间、导致食竹缺乏、隔断大熊猫之间的交流等。人类对大熊猫栖息地空气、水质、土壤的污染，包括外来生物的污染，都对大熊猫的生存构成了巨大威胁。

The main threat of giant panda is from human activities, such as deforestation, farmland reclamation, road and bridge construction, and hydropower station construction. Those human activities will destroy the giant panda's habitats. Thus, the habitats of giant pandas are shrinking, bamboos may be scarce, and communication between giant pandas is cut off. The air, water, soil and exotic organism pollutions in the habitats of giant panda, all pose a great threat to the survival of giant pandas.

卧龙自然保护区于1963年创建，1997年，四川卧龙国家级自然保护区管理局成立，与卧龙特别行政区合署办公。1983年，中国保护大熊猫研究中心成立。现在全国共有以保护大熊猫为主的保护区67个。国家为了强化大熊猫保护，进一步整合四川、陕西、甘肃的自然和人力资源，大熊猫国家公园于2018年10月正式挂牌运行，这将开辟大熊猫物种保护的新里程。

Wolong Natural Reserve was established in 1963. In 1997, the Sichuan Wolong National Natural Reserve Administration was established, that joint office with Sichuan Wenchuan Wolong Special Adminitrative Region. In 1983, the China Conservation and Research Center for Giant Panda was established. At present, there are 67 natural reserves be established for giant panda conservation. China officially established the Giant Panda National Park in October 2018, which will integrate natural and human resources of Sichuan, Shannxi and Gansu provinces for strengthen giant panda protection. It is a new milestone in the conservation of giant panda.

中国政府长久以来实施的大熊猫保护工程及《野生动物保护法》的颁布，使偷猎和破坏大熊猫栖息地环境的现象得到了有效遏制。自2000年10月起实施的《天然林资源保护工程》及持续的其他保护措施，可见生态修复效果巨大，生物多样性得到恢复，濒危的野外大熊猫种群数量已达到了1 864只。

The giant panda protection projects and the promulgation of the Wildlife Conservation Law of Chinese government have effectively stopped the poaching and destruction of giant panda habitat. Since October 2000, the implementation of the "Natural Forest Resources Protection Project" and other continuous conservation measures made huge ecological restoration effect, biodiversity has been restoring, the wild population of giant panda has reached 1,864.

中华人民共和国成立后，1953年在都江堰玉堂镇赵公山救助了一只大熊猫，为了使其得到良好的治疗与养护，成都动物园成为第一家饲养大熊猫的动物园。此后，北京、西安、昆明、重庆、福州、上海也陆续开始人工圈养大熊猫，从此开启了大熊猫的迁地保护事业。

After the founding of People's Republic of China, a giant panda was rescued at Zhaogongshan, Yutang, Dujiangyan in 1953. In order to ensure its professional treatment and maintenance, Chengdu Zoo became to be the first zoo to raise giant pandas. Since then, giant pandas have been raised in Beijing, Xi'an, Kunming, Chongqing, Fuzhou and Shanghai, which started the cause of off-site conservation too.

目前，大熊猫的主要圈养保护机构，只有中国保护大熊猫研究中心、成都大熊猫繁育研究基地、北京动物园和陕西省野生动物救护中心四家。通过多家机构广大科技人员四十余年的共同努力，攻克了许多人工圈养大熊猫的技术难题，大熊猫人工饲养繁育技术已经基本成熟，现在人工圈养的大熊猫数量已经达到548只。

Currently, there are only four main captive conservation institutions (China Conservation and Research Center for Giant Pandas, Chengdu Research Base of Giant Panda Breeding, Beijing Zoo and Shaanxi Wildlife Rescue Center). Through more than 40 years of joint efforts, the technical difficulties of captive giant panda breeding were solved. The captive breeding technology of giant panda has been basically mature, and now the number of captive giant panda has reached 548.

与大熊猫生活在同一个生态环境的动物很多，其伙伴主要有金丝猴、小熊猫、牛羚、黑熊、藏酋猴、野猪等动物。因为它们喜食的食物存在差异，生活习性也不同，所以这些野生动物都生存在同一个生态环境下。在生态系统与植物群落完整的情况下，野生动物之间相安无事，均得到很好的生息繁衍。

There are many other animals living in the same habitat with the giant panda, such as golden monkey, red panda, takin, black bear, Tibetan macaque and wild boar. Because of their different food preferences and habits, these wild animals all live in the same ecological environment without competition. When the ecosystem and plant communities are intact, wild animals live and reproduce peacefully.

在大熊猫栖息地，已经不见华南虎的身影，现在几乎没有什么动物能够威胁到成年大熊猫，只有金钱豹、金猫、黄喉貂等几种动物能对幼年期的大熊猫构成威胁。个别地域的狼和流浪的野狗会干扰大熊猫的正常生存，甚至于威胁到大熊猫幼子的生命安全。

Because south China tiger has not been seen in the giant panda habitat area, the adult giant panda has fewer natural enemies. Only the leopard, the golden cat and yellow-throated marten can threaten the infancy giant panda. Wolves and stray dogs in certain areas can interfere giant pandas, even threaten the life of giant panda cubs.

无论是野生大熊猫还是人工圈养的大熊猫，其主要疾病有犬瘟热等传染性疾病、消化道疾病、体内外寄生虫等疾病。现在有关机构正加强大熊猫栖息地周边群众的宣传教育，有效控制人类生产、生活行为，努力降低人畜疾病传播给大熊猫的风险。另外，科技界也正在努力研究预防大熊猫传染性疾病的疫苗，疫苗研制成功后，至少可以保护人工圈养大熊猫免受传染病的威胁。

The main diseases of wild and captive giant panda include: distemper and other infectious diseases, digestive tract diseases, internal and external parasites. Now relevant institutions are strengthening the publicity and education of the people around panda habitat, and control human activities, efforts to reduce the risk of zoonosis to giant pandas. The scientific community is also working on developing vaccines to prevent infectious diseases, which would protect captive giant pandas at least.

友谊的使者

The Messenger of Friendship

大熊猫深受世界人民的喜爱，因此，中国政府于1957~1982年向9个友好国家赠送了23只大熊猫。此后，也向很多友好国家提供大熊猫进行巡展和合作科学研究，给喜爱大熊猫的人们提供了一个亲眼目睹的宝贵机会，受到了世界各国人民的欢迎。

Giant pandas are loved by people all over the world. Therefore, Chinese government presented 23 giant pandas to 9 friendly countries from 1957 to 1982. Since then, giant pandas have also been provided to many friendly countries for exhibitions and cooperative scientific research. This provides a precious opportunity for people who love pandas to witness with their own eyes, which has been welcomed by people all around the world.

据历史记载，唐朝女皇武则天为传达友谊，曾赠送给当时的日本天皇2只大熊猫和70张毛皮。最为轰动的是1972年中美建交之际，中国政府向美国人民赠送的“玲玲”“兴兴”大熊猫国礼，受到美国民众的热烈欢迎，这一举措被称为“熊猫外交”。2019年4月在中俄建交70周年之际，中国政府又提供了一对大熊猫“如意”和“丁丁”前往俄罗斯莫斯科动物园旅居，为“新时代中俄全面战略协作伙伴关系”的提升助了一臂之力。

According to the record, the Emperor Wu Zetian of the Tang Dynasty presented 2 giant pandas and 70 skins to the Emperor of Japan as a token of friendship. The most sensational event was China presented "Ling Ling" and "Xing Xing" as national gifts to the American people in 1972 to celebrate two countries established diplomatic relations. These gifts were warmly welcomed by the American people and were called "panda diplomacy". In April 2019, on the occasion of 70th anniversary of establishing diplomatic relations between China and Russia, Chinese government provided a pair of giant pandas, Ruyi and Dingding, to Moscow zoo in Russia, which is boosting the "comprehensive strategic cooperative partnership between China and Russia " in the new era.

目前，我国已与17个国家、22个动物园开展了保护大熊猫合作研究项目，在外参与国际合作研究项目的大熊猫数量达到58只。现在全球有大熊猫合作研究项目的主要国家有：美国、俄罗斯、西班牙、法国、德国、瑞典、英国、丹麦、芬兰、荷兰、澳大利亚、奥地利、日本、马来西亚、泰国。

At present, China has carried out cooperative research projects of panda conservation with 22 zoos in 17 countries. The number of giant panda participating in international cooperative research projects has reached 58. The countries with giant panda cooperation projects include United States, Russia, Spain, France, Germany, Sweden, the United Kingdom, Denmark, Finland, the Netherlands, Australia, Austria, Japan, Malaysia and Thailand.

都江堰市是一个大熊猫高度聚集的生态城市，在该市管辖区域内，有两家世界顶级大熊猫科研与保护机构，人工圈养着50余只大熊猫，野外也有8只大熊猫分布于龙溪-虹口国家级自然保护区内。通过与国外的交流合作，都江堰市与很多国家的城市建立了友好关系，从而促进了国际间的文化交流和友谊，强化了相互的理解与互惠合作。

Dujiangyan is an ecological city with a high concentration of giant pandas. Within the jurisdiction of the city, there are two world's top research and conservation institutions for giant pandas, and they have more than 50 captive giant pandas. There are 8 giant pandas live at Longxi-Hongkou, Dujiangyan National Natural Reserve wild area. Through the exchange and cooperation with foreign countries, Dujiangyan established friendly relationships with many other foreign cities. Thus, giant pandas promote international cultural exchanges, deepen friendship and strengthen mutual understanding and cooperation.

熊猫智慧 人类借鉴

The Philosophy of Giant Panda

不同地域的生态系统是受大气环境的湿热分布和地理环境影响和制约的。每个地域的生命形式、物种类型、生存方式，其实都顺应了自然规律。在物竞天择的大环境下，大熊猫能够生存到今天，就是自然选择、生存竞争的结果。因此，人类必须敬畏大自然，顺应自然规律才能保持人类社会的可持续发展。

Ecosystems in different regions are affected and restricted by the distribution of heat and humidity in the atmospheric environment and the geographical environment. In fact, the life forms, species types and living styles of each region conform to the natural laws. In the natural selection environment, giant pandas still living is the result of natural selection and life struggling. Therefore, it suggests that human beings must respect nature and conform to the laws of nature to maintain sustainable development of human society.

自然界的生命形式复杂多样，但都顺应着自然规律，遵循着各自的规则生存下来。大熊猫的祖先是食肉动物，估计在当时的竞争中不占优势，由食肉过渡到杂食，再进一步过渡到食竹。竹子为大熊猫的生存提供了稳定的营养来源。大熊猫的这一系列演化，体现了物竞天择，适者生存的自然规律。

Life forms in nature are complex and diverse, but they all adapt to the laws of nature and follow their own rules to survive. The ancestor of giant panda is carnivore, they might not have advantage in competition with other carnivores at that time. So, they changed from carnivorous to omnivorous, and then gradually to staple bamboo. Bamboos provides a steady source of nutrients for giant pandas. The series of evolution of giant panda reflect that creatures only conform to the natural laws can be extant in the survival competition.

动物学里有个生态位的概念，比如，自然界有陆生动物、水生动物，有食肉的和食草的，有夜间觅食与昼间觅食的，以上种种就有了各自的生态位，从而降低了相互之间的竞争，达成了和谐相处、共享资源、共同生存的理想状况。这对在人类生产经济活动中，合理调整产业结构、地域和空间布局，处理好供需关系等都有一定启发。

Niche is a concept in zoology. There are terrestrial animals, aquatic animals, carnivores and herbivores, nocturnal animals and diurnal animals, all of these animals have their own niche, which reduces the competition between each other. In the economic activities of human production, it has certain enlightenment to rational adjustment of industrial structure, regional and spatial distribution, and handle the relationship between supply and demand.

在生物进化过程中，特别是动物界，其器官结构与功能都在经历不断的适应性进化改变，大熊猫的进化处处都印证了用进废退的自然规律。由此启发人们要适度运动，强身健体，使各个器官保持正常运转，为提高工作和生活的质量铺垫好基础；开动脑筋，强化后天的学习，多阅读，吸取各方面知识，提高对自然规律和社会发展规律的认知，这样才能使大脑不迟钝，心态更平和。

In the process of evolution, especially in the animal kingdom, the organ structure and function are constantly undergoing adaptive evolution. The evolution of the giant panda proved the natural law of use and disuse. Therefore, that inspire people to do moderate exercise or physical fitness, making each organ operational normally, for improving the quality of work and life. We also should think more, learn more, read more and absorb all aspects of knowledge to improve the understanding of natural laws and social development. In this way, it can prevent brain retardation, and keep a calm mind.

大熊猫面对低能量食物采取了多种对策，才使生命得到延续。大熊猫提醒人们要倡导低消耗，不要无谓的消耗和浪费资源。提倡人们在工作和生活中厉行节约，节约用纸和不使用一次性筷子等木制品，为保护绿色的森林做一些力所能及的事情。人类每一个个体都要节约资源，特别是不可再生的资源，这样才能使人类社会健康发展。

大熊猫主食竹子，不是掠食性动物，因此没有食肉动物那么凶猛，大熊猫显得温顺。特别是未成年的大熊猫，相互之间非常友善，关系融洽。但受到威胁时，大熊猫力大无比，它的咬合力和抓力出众，一般动物都很难抵抗大熊猫的强力反击。其实，大熊猫具有外柔内刚的性格，也符合中国人的传统性格与理念，那就是和谐相处，先礼后兵。

Giant pandas take a variety of countermeasures to survive for their low energy diet. That reminds people to advocate low consumption without wasting, encourage people to practice thrift in work and life, save paper and do not use disposable chopsticks or other disposable wood products. That can help us protecting our forests. Everyone should save resources, especially the non-renewable resources. Only in this way that can make our human society develop healthy.

Giant panda is not a predator. Bamboo is their main food source. Unlike other raptorial carnivores, giant panda looks very meek. Especially the cubs, they are harmonious to each other. However, giant panda has great power to prevent threats from other animals. Most animals cannot bear their strong biting and seizing force. In fact, the traditional concept and character of Chinese people are just like giant panda, outwardly yielding but inwardly firm, advocating harmonious coexistence and try fair means before resorting to force.

图书在版编目（CIP）数据

简说大熊猫：汉英对照 / 龙溪-虹口国家级自然保护区管理局编著. -- 成都：四川科学技术出版社, 2019.12（2021.6重印）

ISBN 978-7-5364-9675-0

Ⅰ. ①简… Ⅱ. ①龙… Ⅲ. ①大熊猫—青少年读物—汉、英 Ⅳ. ①Q959.838-49

中国版本图书馆CIP数据核字(2019)第254848号

简说大熊猫

Jianshuo Daxiongmao

龙溪—虹口国家级自然保护区管理局　编著

出品人	程佳月
责任编辑	程蓉伟
出版发行	四川科学技术出版社
封面设计	李　庆
责任印制	欧晓春
设计制作	四川蓝色印象艺术设计有限公司
印　　刷	四川省南方印务有限公司
成品尺寸	215mm × 203mm
印　　张	6
字　　数	20千
版　　次	2019年11月第1版
印　　次	2021年6月第3次印刷
书　　号	ISBN 978-7-5364-9675-0
定　　价	45.00元